Acid, Acid Everywhere

Exploring Chemical, Ecological, and Transportation Systems

Second Edition

A Problem–Based Learning Unit Designed
for 4th–6th Grade Learners

KENDALL/HUNT PUBLISHING COMPANY
4050 Westmark Drive   Dubuque, Iowa   52002

Book Team
Chairman and Chief Executive Officer: *Mark C. Falb*
President and Chief Operating Officer: *Chad M. Chandlee*
Director of National Book Program: *Paul B. Carty*
Editorial Development Manager: *Georgia Botsford*
Developmental Editor: *Lynnette M. Rogers*
Vice President, Operations: *Timothy J. Beitzel*
Assistant Vice President, Production Services: *Christine E. O'Brien*
Senior Production Editor: *Charmayne McMurray*
Permissions Editor: *Renae Horstman*
Cover Designer: *Jenifer Chapman*

Author Information for Correspondence and Workshops:
Center for Gifted Education
The College of William and Mary
P.O. Box 8795
Williamsburg, VA 23187-8795
Phone: 757-221-2362
Email address: *cfge@wm.edu*
Web address: *www.cfge.wm.edu*

Center for Gifted Education Staff, First Edition
Project Director: Dr. Joyce Van Tassel-Baska
Project Managers: Dr. Shelagh A. Gallagher
Dr. Victoria B. Damiani
Project Consultants: Dr. Beverly T. Sher
Linda Neal Boyce
Dana T. Johnson
Dr. Donna L. Poland
Teacher Developer: Sandra Coleman

Center for Gifted Education Staff, Second Edition
Executive Director: Dr. Joyce Van Tassel-Baska
Director: Dr. Elissa F. Brown
Curriculum Director and Unit Editor: Dr. Kimberley L. Chandler
Curriculum Writers: Brandy L. Evans
Susan McGowan
Curriculum Reviewers: Dr. Beverly T. Sher
Sharon Bowers

The William and Mary Center for Gifted Education logo is a depiction of the Crim Dell Bridge, a popular site on the William and Mary campus. Since 1964 this Asian-inspired structure has been a place for quiet reflection as well as social connections. The bridge represents the goals of the Center for Gifted Education: to link theory and practice, to connect gifted students to effective learning experiences, to offer career pathways for graduate students, and to bridge the span between general education and the education of gifted learners.

Contents

Glossary

Abiotic Element Any nonliving component of an ecosystem such as light, water, nutrients, or air

Assumption Conclusion based on one's beliefs and presuppositions

Acidic Solution A water-based solution containing more hydronium than hydroxyl ions; on a pH scale, these solutions have a pH value lower than 7

Basic Solution A water-based solution containing more hydroxyl ions than hydronium ions; on a pH scale, these solutions have a pH value greater than 7

Bias A one-sided or slanted view that may be based on culture, experience, or other aspects of one's background

Biotic Element Any living component of an ecosystem such as a microbe, a deer, or a human being

Boundary Something that indicates a border or limit

Constant The factor(s) in an experiment that remain the same

Control The standard of comparison in an experiment

Corrosive Causing gradual destruction

Dependent Variable The variable that may be influenced by the independent variable

Dilution The process through which a solution's pH is brought closer to 7 by adding water

Dysfunctional Not working properly

Element Essential component of a system

Hypothesis A tentative explanation for an observation, phenomenon, or scientific problem that can be tested by further investigation

Implication A suggestion of likely or logical consequence; a logical relationship between two linked propositions or statements

Independent Variable The variable that is changed intentionally by the experimenter

Inference Interpretation based on observation

Input Something put into a system

Interaction Mutual action or influence of one element, individual, or group with another; a process occurring among elements of a system

Interdependent Mutually relying on or requiring the aid of another

Mean The average value of a set of numbers

Median The middle value in a distribution of numbers

Mode The value or item occurring most frequently in a series of data

Neutralization The process by which an acidic (or basic) solution is brought to a pH of 7 by adding a base (or an acid)

Output A product of the system interactions

Perspective An attitude, opinion, or position from which a person understands a situation or issue

pH A measure of the acidity or alkalinity of an aqueous solution

pH Neutral An equal number of hydroxyl and hydronium ions represented by a pH value of 7 on the pH scale

pH Indicators Strips of specially treated paper that change colors when dipped in an aqueous solution and can be used to determine a solution's pH

Productive Effective in achieving specified results

Property A characteristic or attribute

Range The difference between the smallest and largest values in a frequency distribution

Reasoning Evidence or arguments used in thinking

Stakeholder An individual with an interest in or involvement with an issue and its potential outcomes

System A group of interactive, interrelated, or interdependent elements that form a complex whole

Titration A method for determining the concentration of an unknown acid or base in a solution

Topographic Graphic representation of the surface features of a place or region on a map, indicating their relative positions and elevations

Validity Accuracy and generalizability of the experiment's outcomes

Laboratory Safety Precautions

As this unit involves laboratory work, some general safety procedures must be observed at all times. Most school districts have prescribed laboratory safety rules; for those that do not, some basic rules to follow for scientific experimentation include:

1. Appropriate behavior in the lab includes no running or horseplay; materials should be used for intended purposes only.
2. Eating, drinking, or smoking is prohibited in the lab; there should be no tasting of laboratory materials.
3. If students are using heat sources, such as alcohol burners, long hair must be tied back and loose clothing should be covered by a lab coat. Long sleeves should be rolled or pushed up on the arms.
4. Fire extinguishers must be available; students should know where they are and how to use them.

Specific safety rules relevant to implementing this unit include:

1. This unit covers acid/base chemistry. Use of concentrated HCl, dangerous household acids and bases such as bleach or oven cleaner should not be allowed. Any household item that carries a warning label should be considered off limits for student use although they should be encouraged to look at the warning labels on such products in their study of acids and bases. Food-type acids and bases such as vinegar, lemon juice, and baking soda are safe for student use; moreover, use of such acids and bases does not require gloves or a lab apron.
2. Safety goggles should be worn at all times in the lab; some provision for washing chemicals out of eyes should be made. Either a sink with an eyewash attachment or a large squirt bottle full of water should be available.
3. When diluting a concentrated acid or base, add the concentrated acid or base to water slowly to minimize the likelihood of splashing the acid or base.
4. Do not mix concentrated acids with concentrated bases, as the resulting reaction may be vigorous and dangerous. Do not mix dangerous household chemicals.

_____   _____
Student Signature   Date

_____   _____
Teacher Signature   Date

Content Pre-Assessment

1. a. Draw the pH scale and label its parts.

   b. On your pH scale drawing, show where the following substances belong: highly acidic substances, neutral substances, and highly basic substances.

2. How would you change the pH of your clear liquid from acidic to neutral? List the two different ways to do this.

   a.

   b.

Experimental Design Pre-Assessment

Construct a fair test of the following question: *Is orange juice an appropriate food for plants?*

Describe in detail how you would test this question. Be as scientific as you can as you write about your test. Write the steps you would take to find out if orange juice is an appropriate food for plants.

Adapted from Fowler, M. (1990). The diet cola test. *Science Scope, 13(4),* 32–34.

Systems Pre-Assessment

Think of how a swimming pool is a system.

1. List the parts of the system in the spaces provided below. Include boundaries, elements, input, and output.

   Boundaries:

   Elements:

   Inputs:

   Outputs:

2. Draw a diagram of the system showing where each of the parts is located.

Name _____   Date _____

Systems Diagram

Name of System _____

Boundaries

Elements

Inputs

Outputs

Interactions

Name _____ Date _____

Systems Model

Name of System _____

Give examples of how the system demonstrates each generalization.

The interactions and outputs of a system change when its inputs,
elements, or boundaries change.

Systems can be productive or dysfunctional.

Many systems are made up of smaller systems.

Systems are interdependent.

All systems have patterns.

Initial Problem Statement*

You are the supervisor of the day shift of the State Highway Patrol. It is 6:00 AM on a cool autumn morning. You are sleeping when the phone rings. You answer and hear, "Come to the _Clear Creek Bridge on Route 15_. There has been a major accident and you are needed."

Quickly you dress and get on the road to hurry to the site of the emergency. As you approach the bridge, you see an overturned truck that has apparently crashed through a metal guard rail. The truck is missing a wheel and is perched on its front axle. You see "CORROSIVE" written on a small sign on the rear of the truck. There is a huge gash in the side of the truck and from the gash a liquid is running down the side of the truck, onto the road, and down the hill into a creek. Steam is rising from the creek. All traffic has been stopped and everyone has been told to remain in their cars. Many of the motorists trapped in the traffic jam appear to be angry and frustrated. Police officers, firefighters, and rescue squad workers are at the scene. They are all wearing coveralls and masks. The rescue squad is putting the unconscious driver of the truck onto a stretcher. Everyone seems hurried and anxious.

*The underlined section indicates the part of the letter which should be tailored to your particular location.

Need to Know Board

What we know . . .	What we need to know . . .	How we can find out . . .

Need to Know Board

What we know . . .	What we need to know . . .	How we can find out . . .

Problem Log Questions

Directions: Paraphrase the problem situation in your Problem Log and then answer the following questions.

1. After our first two days of discussion, what do you think the situation really is?

2. Why do you think this is the main problem?

3. How has your assumption changed since we first started talking?

continued

4. How is this situation an example of a complex system?

5. What issues are you most interested in investigating?

Smudged Invoice

Invoice #3944

Coleman Chemicals

Account # 2231980	Sold By ACA	Reference # 2284	Ship Via Jones Trucking	Terms Net 30	Date 1-15-07

Qty. Ordered	Qty. Shipped	Item #	Description	Unit Price	Total Price
5000 gal.	5000 gal.	6543	XG147B	3.00	$15,000.00

Sale Amount	$15,000.00
Sales Tax	$ 750.00
Freight	$ 2,000.00
Total	$17,750.00

Please enter our order for the above merchandise subject to the terms
and conditions on the reverse side of this order.

Reasoning Wheel

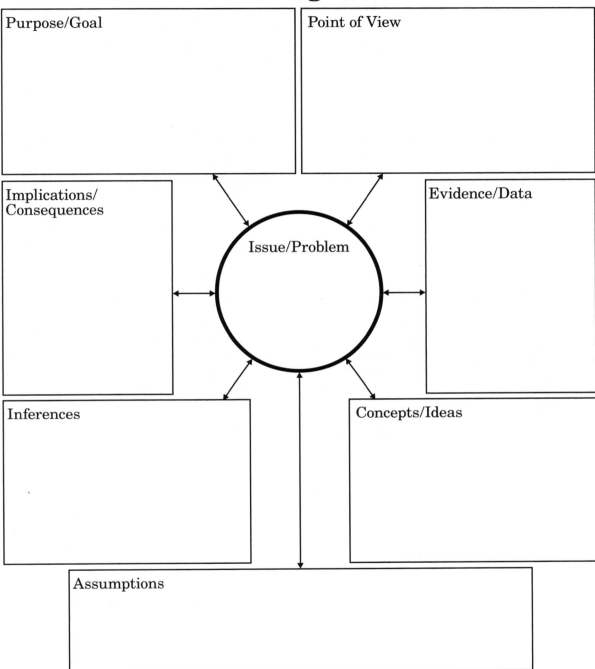

| Purpose/Goal | Point of View |

Implications/Consequences

Issue/Problem

Evidence/Data

Inferences

Concepts/Ideas

Assumptions

Adapted from Paul, R. (1992). *Critical thinking: What every person needs to survive in a rapidly changing world.* Sonoma, CA: Foundation for Critical Thinking.

Name _____   Date _____

Traffic Detour Report

Name of Highway Patrol Officer: _____

Date of Report: _____

Please consult with Highway Patrol Officers and make a detour route recommendation for the flow of traffic around the accident. Describe the exact route of your detour then complete the rest of this report. Be prepared to defend your decision.

1. What are the positive aspects of your detour?

2. What are the negative aspects of your detour?

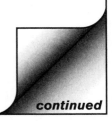

continued

3. Which stakeholders did you consider when creating your detour?

4. What assumptions did you make about the stakeholders whom you did not accommodate?

5. How does traffic relate to the concept of systems?

Name _____   Date _____

Need to Know Board

What we know . . .	What we need to know . . .	How we can find out . . .

Weather Report

After the early morning fog burns off, today will be clear, with winds from the west at 5–10 mph. The high will be about 65 degrees, with the overnight low between 55 and 60 degrees. Tomorrow's high will be 65 degrees, with wind from the west at 20 to 25 mph.

Hazard Alert Form

Highway Patrol Officer (Name): _____

Date of report: _____

Specific hazard posted: _____

Site(s) which need to be posted:

_____

_____

_____

_____

Reasons for posting site:

_____

_____

_____

_____

Who needs to be informed? Why? _____

_____

_____

_____

What are the possible effects of the hazard?

_____

_____

_____

Name _____   Date _____

Student Brainstorming Guide

1. What is the scientific problem?

2. What materials are available to help define a solution to the problem?

continued

31

3. How would these materials best be used in designing a solution to the problem?

4. What hypothesis can be formed before conducting this experiment?

5. What must be observed or measured in order to find an answer to the scientific question?

Experimental Design Diagram

Title:

_____

Hypothesis:

_____

Independent Variable:

_____

Dependent Variable:

_____

Observations/Measurements:

_____

Constants:

_____

Control:

_____

Safety Procedures:

Name _____   Date _____

Experimental Design Diagram Checklist

1. _____ Does the *title* clearly identify both the independent and dependent variables?

2. _____ Does the *hypothesis* clearly state how you think changing the independent variable will affect the dependent variable?

3. _____ Is there just one clearly defined *independent variable?*

4. _____ Is the *dependent variable* clearly stated?

5. _____ Is the expected response of the dependent variable clearly stated?

6. _____ Do the *observations / measurements* to be made include appropriate units of measure?

7. _____ Are all *constants* listed and clearly defined?

8. _____ Is there a clearly stated *control?*

9. _____ Is the experiment thoughtfully designed?

10. _____ Are safety procedures taken into consideration?

Experimental Protocol

1. From the available materials, list the materials needed for the experiment:

2. Write a step-by-step description of the experiment. List every action that must be taken during the experiment.

3. Design a data table to collect and analyze your information.

Name _____   Date _____

Laboratory Report

1. What did you do or test?

2. How did you do it? What materials and methods did you use?

3. What did you find out? (Include a data summary and the explanation of its meaning.)

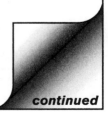

continued

4. What did you learn from your experiment?

5. What additional questions do you have?

6. How does the information you learned help with the problem?

7. Was your experimental setup a system? Why or why not?

Problem Log Questions

1. Why would it be important to know the flow rates of the creek and of the dripping liquid in the problem?

2. Given the flow rate that was calculated from the dripping faucet:
   a. How much water flows from the faucet every minute?

   b. How much water flows from the faucet every day?

   c. How much water would flow from the faucet in a year?

continued

3. How do weather and climate affect flow rate? How might weather, climate, and the season affect our problem?

pH Scale

Concentration of hydrogen ions compared to distilled water

10,000,000	pH = 0
1,000,000	pH = 1
100,000	pH = 2
10,000	pH = 3
1,000	pH = 4
100	pH = 5
10	pH = 6
1	pH = 7
1/10	pH = 8
1/100	pH = 9
1/1000	pH = 10
1/10,000	pH = 11
1/100,000	pH = 12
1/1,000,000	pH = 13
1/10,000,000	pH = 14

pH Data Collection

This pH chart may be used to record initial results of your experiment.

Concentration of hydrogen ions compared to distilled

	water	Substance
10,000,000	pH = 0	
1,000,000	pH = 1	
100,000	pH = 2	
10,000	pH = 3	
1,000	pH = 4	
100	pH = 5	
10	pH = 6	
1	pH = 7	
1/10	pH = 8	
1/100	pH = 9	
1/1000	pH = 10	
1/10,000	pH = 11	
1/100,000	pH = 12	
1/1,000,000	pH = 13	
1/10,000,000	pH = 14	

Problem Log Questions

1. Describe in your own words what the pH scale measures.

2. What was the *most significant* outcome of today's experiments?

3. Is this knowledge in any way useful in the investigation of the spill? Why? How might it be used?

4. List and describe the safety procedures discussed in class today. Include both safety equipment and safe behaviors.

5. After today's activity, do you think we need to add more to the list of safe behaviors? Why or why not? If so, what are they?

6. Are any of our rules unnecessary? If so, which ones and why?

Name _____   Date _____

County Commissioner Letter*

Dear Students of Ms./Mr. _____,

I have heard from the principal of your school that you are interested in the events which occurred on the highway just over <u>Clear Creek</u> last <u>Wednesday</u>. When I was a small boy/girl growing up in the county, science fascinated me; however, I chose to study law when I entered college.

I am eager to support you in your efforts to find a solution to the problems at <u>Clear Creek</u>, so I have provided a small sample of the spill for you to test. If you discover a cost-saving way to keep the ecosystem in <u>Clear Creek</u> from deteriorating, I would be most anxious to hear from you.

Please tell your family to remember me when they go to vote next fall!

Sincerely,

Commissioner Gordon
<u>Curry County</u>

*Underlined sections indicate parts of the letter which should be tailored to your particular location.

Name _____ Date _____

Need to Know Board

What we know . . .	What we need to know . . .	How we can find out . . .

Name _____   Date _____

Graph Paper

Graph Title: _____

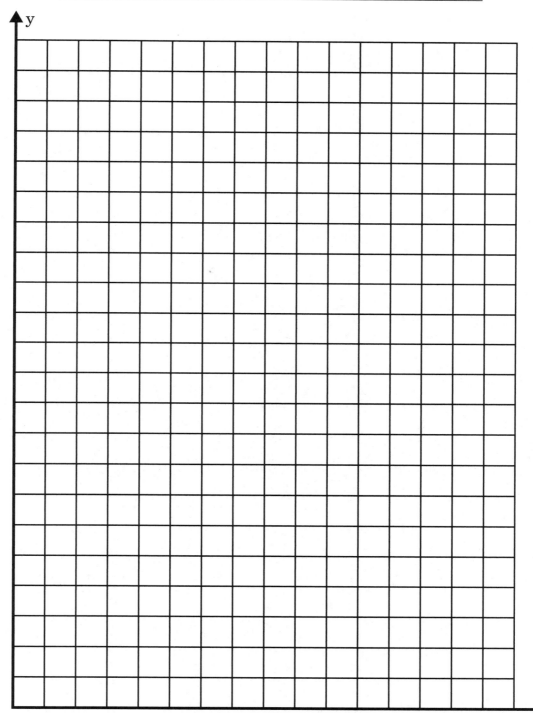

Need to Know Board

What we know . . .	What we need to know . . .	How we can find out . . .

Name _____   Date _____

Student Brainstorming Guide

1. What is the scientific problem?

2. What materials are available to help define a solution to the problem?

continued

3. How would these materials best be used in designing a solution to the problem?

4. What hypothesis can be formed before conducting this experiment?

5. What must be observed or measured in order to find an answer to the scientific question?

Experimental Design Diagram

Title:
_____

Hypothesis:
_____

Independent Variable:
_____

Dependent Variable:
_____

Observations/Measurements:

_____

Constants:
_____

Control:
_____

Safety Procedures:

Name _____   Date _____

Experimental Design Diagram Checklist

1. _____ Does the *title* clearly identify both the independent and dependent variables?

2. _____ Does the *hypothesis* clearly state how you think changing the independent variable will affect the dependent variable?

3. _____ Is there just one clearly defined *independent variable?*

4. _____ Is the *dependent variable* clearly stated?

5. _____ Is the expected response of the dependent variable clearly stated?

6. _____ Do the *observations / measurements* to be made include appropriate units of measure?

7. _____ Are all *constants* listed and clearly defined?

8. _____ Is there a clearly stated *control?*

9. _____ Is the experiment thoughtfully designed?

10. _____ Are safety procedures taken into consideration?

Experimental Protocol

1. From the available materials, list the materials needed for the experiment:

2. Write a step-by-step description of the experiment. List every action that must be taken during the experiment.

3. Design a data table to collect and analyze your information.

Name _____ Date _____

Student Brainstorming Guide

1. What is the scientific problem?

2. What materials are available to help define a solution to the problem?

3. How would these materials best be used in designing a solution to the problem?

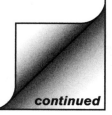

continued

4. What hypothesis can be formed before conducting this experiment?

5. What must be observed or measured in order to find an answer to the scientific question?

Experimental Design Diagram

Title:

_____

Hypothesis:

_____

Independent Variable:

_____

Dependent Variable:

_____

Observations/Measurements:

_____

Constants:

_____

Control:

_____

Safety Procedures:

Name _____   Date _____

Experimental Design Diagram Checklist

1. _____ Does the *title* clearly identify both the independent and dependent variables?

2. _____ Does the *hypothesis* clearly state how you think changing the independent variable will affect the dependent variable?

3. _____ Is there just one clearly defined *independent variable?*

4. _____ Is the *dependent variable* clearly stated?

5. _____ Is the expected response of the dependent variable clearly stated?

6. _____ Do the *observations/measurement*s to be made include appropriate units of measure?

7. _____ Are all *constants* listed and clearly defined?

8. _____ Is there a clearly stated *control?*

9. _____ Is the experiment thoughtfully designed?

10. _____ Are safety procedures taken into consideration?

Experimental Protocol

1. From the available materials, list the materials needed for the experiment:

2. Write a step-by-step description of the experiment. List every action that must be taken during the experiment.

3. Design a data table to collect and analyze your information.

Name _____   Date _____

Laboratory Report

1. What did you do or test?

2. How did you do it? What materials and methods did you use?

3. What did you find out? (Include a data summary and the explanation of its meaning.)

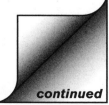

continued

4. What did you learn from your experiment?

5. What additional questions do you have?

6. How does the information you learned help with the problem?

7. Was your experimental setup a system? Why or why not?

Name _____   Date _____

Problem Log Questions

1. Draw a picture of your setup and label its components.

2. What were its boundaries?

3. What were the elements of your experimental setup?

continued

4. Did you put anything into your system?

5. What was the output from your experimental system?

6. How did the elements of your experimental system interact with any input you added? How did the input change your experimental system?

7. What interactions inside the system allowed the system to produce output?

Name _____ Date _____

Systems Diagram

Name of system: _____

Boundaries

Elements

Inputs

Outputs

Interactions

Letter from the Friends of Clear Creek*

To: State Highway Patrol Supervisor

From: Public Works Central Office

An ecological awareness group calling itself _The Friends of Clear Creek_ has sent us the enclosed facsimile. They claim that if the corrosive spill is not immediately addressed, they will alert the local and national media and start a boycott of _Coleman Chemicals_. They are also initiating a lawsuit against the state for improperly handling the spill and another lawsuit against _Coleman Chemicals_ for transporting materials of an acidic nature on a public highway.

Please inform us of the status of your cleanup efforts immediately.

Facsimile:

ALERT! ALERT! ALERT! ALERT!
Clear Creek is getting more and more polluted by the second . . . the water is becoming more dangerous for all living things. No one is taking action to halt the spread of the acid spill. Call your elected representatives. Call the highway patrol. Let them know how you feel about dangerous materials being transported on your highways. Put the pressure on and SAVE _CLEAR CREEK!_

Signed, _The Friends of Clear Creek_

*Underlined sections indicate parts of the letter which should be tailored to your particular location.

Reasoning About a Situation or Event

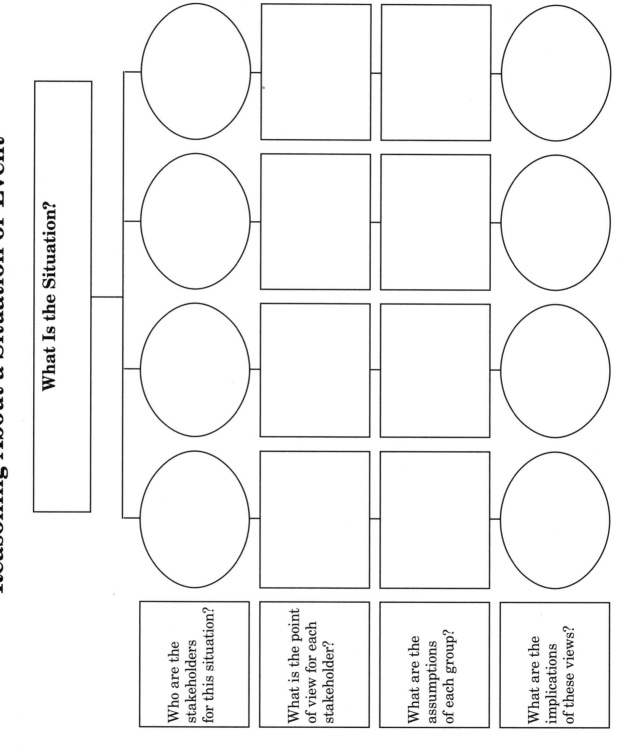

What Is the Situation?

Who are the stakeholders for this situation?

What is the point of view for each stakeholder?

What are the assumptions of each group?

What are the implications of these views?

Problem Log Questions

1. Describe the most important organism in the set of organisms you researched. How does it survive from day to day? What makes it an important part of the ecosystem?

2. How could you find out whether this organism would be harmed directly by the addition of acid to the creek? Would it be ethical to do an experiment to find out? Why or why not?

3. What might be a long-term effect on this organism? On the ecosystem?

Reasoning Wheel

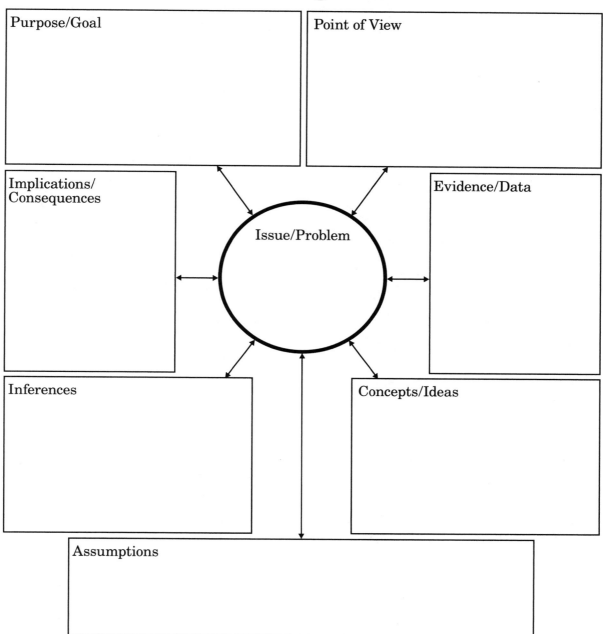

Adapted from Paul, R. (1992). *Critical thinking: What every person needs to survive in a rapidly changing world.* Sonoma, CA: Foundation for Critical Thinking.

Name _____   Date _____

Student Brainstorming Guide

1. What is the scientific problem?

2. What materials are available to help define a solution to the problem?

3. How would these materials best be used in designing a solution to the problem?

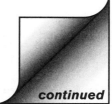

continued

4. What hypothesis can be formed before conducting this experiment?

5. What must be observed or measured in order to find an answer to the scientific question?

Experimental Design Diagram

Title:

_____

Hypothesis:

_____

Independent Variable:

_____

Dependent Variable:

_____

Observations/Measurements:

_____

Constants:

_____

Control:

_____

Safety Procedures:

Experimental Design Diagram Checklist

1. _____ Does the *title* clearly identify both the independent and dependent variables?

2. _____ Does the *hypothesis* clearly state how you think changing the independent variable will affect the dependent variable?

3. _____ Is there just one clearly defined *independent variable?*

4. _____ Is the *dependent variable* clearly stated?

5. _____ Is the expected response of the dependent variable clearly stated?

6. _____ Do the *observations/measurements* to be made include appropriate units of measure?

7. _____ Are all *constants* listed and clearly defined?

8. _____ Is there a clearly stated *control?*

9. _____ Is the experiment thoughtfully designed?

10. _____ Are safety procedures taken into consideration?

Name   Date

Experimental Protocol

1. From the available materials, list the materials needed for the experiment:

2. Write a step-by-step description of the experiment. List every action that must be taken during the experiment.

continued

3. What data will be generated from the experiment?

4. Design a data table to collect and analyze your information.

Name _____   Date _____

Laboratory Report

1. What did you do or test?

2. How did you do it? What materials and methods did you use?

3. What did you find out? (Include a data summary and the explanation of its meaning.)

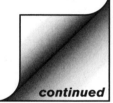

continued

4. What did you learn from your experiment?

5. What additional questions do you have?

6. How does the information you learned help with the problem?

7. Was your experimental setup a system? Why or why not?

Name _____   Date _____

Problem Log Questions

1. What effect did the acid have on plants?

2. How does this new information affect the problem?

3. Could you test the effects of acid on other kinds of organisms? Choose an organism and describe a way of finding out how the organism would react to acid.

4. Would it be ethical to conduct an experiment on your chosen organism? Why or why not?

Name _____   Date _____

Need to Know Board

What we know . . .	What we need to know . . .	How we can find out . . .

Need to Know Board

What we know . . .	What we need to know . . .	How we can find out . . .

Visitor Planning Sheet

Student Name _____

Name of Visitor _____

Visitor's Title _____

What is the visitor's area of expertise?

_____

_____

How/why might the visitor be relevant to a potential problem statement solution?

_____

_____

_____

What would you like to tell the visitor about the problem statement and our work thus far?

_____

_____

_____

What questions are important to ask the visitor?

_____

_____

_____

St. Louis Post-Dispatch Article

February 8, 2004

DANGEROUS CARGO ON OUR ROADS, RAILS by Ken Leiser of The Post-Dispatch

First came the early morning rap on the door. Then came the coughing, the burning eyes.

In the frantic moments that followed a May 17, 2003, hydrochloric acid spill on nearby U.S. Highway 61, Shorti Garner and her husband, Steve, woke their children and piled them into the family camper to flee their home.

"My children—in blankets and all—I scooped them up," Shorti Garner said.

There have been no lingering health effects in the Garner household since 2,500 gallons of hydrochloric acid spilled from an overturned tanker about a quarter-mile away that morning, although Garner no longer lets her children play in the creek that runs through the back yard.

It's a scene that plays out more than once a day, on average, somewhere in the United States. Last year, there were at least 400 transportation accidents like the one in Palmyra in which releases of hazardous materials shut down major transport arteries, displaced people from their homes or businesses, or even resulted in deaths.

Nobody is immune from a possible knock on the door. Major transportation accidents happen in small towns and major cities—almost anywhere there's an active freight line, highway, river or airstrip.

Just last month, four people died when a tractor-trailer loaded with gasoline plunged from a highway overpass onto Interstate 95 near Baltimore and burst into flames. The accident shut down the major traffic corridor for hours. One year ago in Tamaroa, Ill., about 1,000 people were evacuated for as long as four days after 21 train cars, some carrying toxic chemicals, derailed. Some burned.

"Somewhere in the United States, every day, you have truck wrecks with all kinds of stuff on it," said Palmyra Fire Chief Chuck Hoehne, who was in charge of the emergency response to the hydrochloric acid spill in May.

Although transport of spent nuclear fuel gets most of the media and law enforcement attention, more than 300 million shipments of hazardous materials crisscross the country each year, moving almost invisibly through communities on interstates, railroads and airplanes.

continued

For many shipments—including tankers filled with the gasoline that fuels our cars, the chlorine used at water treatment plants, and anhydrous ammonia used in farming— often the only giveaway is the diamond-shaped placard that tells what class of dangerous material is inside.

How safely hazardous materials are shipped looms large in Illinois and Missouri, whose geographic locations and confluence of railroads, rivers and interstate highways pull in a significant share of dangerous cargo, transportation officials say.

The number of total spills and other reportable incidents involving hazardous material shipments reached 14,661 last year—roughly double the typical year in the late 1980s, according to preliminary counts by the Department of Transportation's Research and Special Programs Administration.

That's partly because there's more hazardous cargo being shipped, but also because of stricter reporting requirements, a RSPA spokesman said. Reportable spills can be as small as a stain on a package or a puddle of fuel on a gas station parking lot after a tanker delivery.

"It's everywhere," said Alan Roberts, president of the Dangerous Goods Advisory Council. "This is what keeps the country running. The problem is that nobody wants it in their back yard."

Few people besides the driver and the shipper probably knew how close the tanker truck filled with hydrochloric acid was to their back yards as it approached Palmyra the morning of May 17.

This stretch of U.S. Highway 61 is part of the Avenue of the Saints, a mostly north-south route between St. Louis and St. Paul, Minn., that has seen a steady growth in commercial truck traffic in recent years, including shipments of hazardous materials.

On that morning, a red 2003 International hauling a tank of hydrochloric acid was northbound on the four-lane rural highway. The driver told troopers he nodded off and lost control of his truck, which clipped a road sign and separated from the tanker trailer.

The damaged tanker came to rest in the scruffy, grass median near the Highway 168 exit.

Accident reports show that hundreds of gallons of acid spilled into a drainage ditch that dumps into the North River less than a mile away, leaving a trail of dead fish in its path.

When a vapor cloud formed, several families were evacuated.

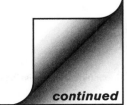

continued

The firefighter explained that a truck had tipped over on the highway and the family should evacuate to a shelter in town.

The federal government regulates the flow of hazardous cargo across America mostly through rules on packaging, preparation and documentation. Shippers must be certified. Transport workers must be trained. Papers must be in order.

Truck driver Greg Pasterski of Wisconsin said he has been driving for 26 years and has never had a ticket or an accident. Pasterski said his company, Air Products, hires only safe drivers.

His trucking company closely examines the trucks before they leave the yard, he said, but government inspectors occasionally have found minor problems out on the road. Mechanics usually "take care of it right away" when they return to the yard.

Still, in some corners of post-Sept. 11 America, environmentalists and community officials are pushing for tougher regulations because of the potential for sabotage or terrorism.

In 2002, Missouri Gov. Bob Holden signed a sweeping counterterrorism measure that, among other things, makes it a Class B misdemeanor for hazardous-materials trucks to enter a highway tunnel. That will only affect trucks using the future Lindbergh Boulevard tunnel beneath the new runway at Lambert Field.

Last month, the District of Columbia city council heard arguments for and against banning shipments of poisonous gases, explosives, and highly flammable gases in the nation's capital, one of the cities considered a "high threat" target for terrorism. Fred Millar, a consultant working for Friends of the Earth on the issue, fears terrorists may attack one of the slow-moving shipments in the future.

"All we are trying to do is reroute the most dangerous targets," Millar said.

Millar said Washington is not the only city at risk. St. Louis and other population centers have shipments of dangerous cargo quietly rumbling through as well, and some would move to regulate them if they knew what they were up against.

"It is very clear that there are no regulations on the routing of hazardous materials by rail, and there is no law or regulation on what the cities or states can do to protect themselves from terrorism," he said.

Federal regulations bar trucks carrying any amount of materials that are highly explosive, poisonous when inhaled or radioactive in heavily populated areas, tunnels or narrow alleys "whenever practical," said David Longo of the Federal Motor Carrier Safety Administration.

The rules also cover larger quantities of other types of dangerous materials, Longo said.

continued

Millar said that regulation is "virtually never enforced."

Roberts, however, said states are required to adopt and enforce the federal routing requirements in order to be eligible for certain highway safety funds.

Joe Delcambre, spokesman for the Department of Transportation's Research and Special Programs Administration, defended the federal government's regulation of hazardous materials shipments, saying safety is the "biggest concern."

Total incidents have grown with the mounting volume of hazardous shipments, including those carried through burgeoning overnight-delivery companies like FedEx and United Parcel Service.

"Those guys really started to take off," Delcambre said. "With an increase in business also came an increase in the number of hazardous material packages."

But Delcambre added that the number of serious transportation mishaps leading to evacuations, highway closures or major injuries has remained fairly constant over the years.

Nuclear shipments

Nearly two months after the acid spill on U.S. Highway 61, another more closely watched shipment passed through Marion County, about 14 miles south of Palmyra on the Norfolk Southern Railroad line.

In that instance, fire and police agencies throughout Marion and surrounding counties were given a near blow-by-blow account of how the shipment of 125 used nuclear fuel rod assemblies from West Valley, N.Y., was progressing as it passed safely.

Transportation routes are studied before nuclear shipments are made. The containers are designed to withstand enormous forces without releasing harmful doses of radiation. They're escorted 24 hours a day, including armed guards near major population centers. And they're generally diverted onto beltways like Interstates 270 and 255 in St. Louis to bypass major urban areas.

Shipments of used nuclear fuel are far less commonplace than other potentially dangerous commodities, numbering a couple dozen a year lately, but they spark the most intense political firefights.

For instance, Holden ordered a three-truck convoy hauling nuclear waste to stop in the Metro East area for several hours in June 2001 before finally allowing it to enter Missouri after rush-hour traffic in St. Louis had dissipated.

continued

With President George W. Bush's administration pushing a nuclear waste repository in Yucca Mountain in Nevada, the transportation issue isn't expected to fade away.

John S. Hark, chairman of Marion County's local emergency planning committee, said that clearly "somebody needed to be notifying somebody" when the latest highly radioactive load passed through Illinois and Missouri between July 13 and 17. "Was there a little more hype than what was needed? Possibly so," Hark said. "But there again, I'd rather there be a little more hype than nothing."

Roberts of the Dangerous Goods Advisory Council said the regulatory safeguards that go into shipping high-level radioactive waste are "very intense." The result? In decades of transporting high-level radioactive waste, he said, there has not been a recorded harmful release of radiation leading to injury or death.

Congress' investigative arm, the General Accounting Office, has concluded that there is a low likelihood of widespread public harm in the event of a terrorist attack or a major accident involving spent nuclear fuel.

But Bob Halstead, a transportation consultant to the Nevada Agency for Nuclear Projects, said it's right for the government to have tougher regulations on used nuclear fuel than any other dangerous chemicals.

Spent nuclear fuel is "immediately lethal" to those who come in direct contact with it, and small amounts of radiation even penetrate its shipment containers, he said. Cleaning up after a worst-case transportation accident could cost up to $10 billion.

"If someone says they have something more dangerous," he said, "let them put it on the table."

Does Halstead believe the rules for transporting other dangerous materials are up to snuff? "You bet I don't," he said.

He thinks it's odd that the government doesn't regulate gasoline or propane more strictly. Drivers should meet tougher training requirements, he says, and the government should take a closer look at routes.

Alerts are impractical

A Post-Dispatch survey of fire departments in the St. Louis area showed that most don't know exactly what is passing through their communities at any given time, although some wish they did.

Officials from 31 fire departments that participated said they realize hazardous materials routinely pass within a half-mile of homes, schools and shopping centers in their communities every day with little or no advance notice.

So many potentially dangerous shipments pass through the region that it would be impractical to alert departments ahead of time, as is done with radioactive materials, some acknowledged.

continued

For fire agencies, that means training and planning for potential disasters involving dangerous chemical spills.

The O'Fallon, Mo., Fire Protection District is crisscrossed by highways and freight railroad tracks. Last year, the district gathered together several local leaders to discuss what would happen if the community were confronted by a railroad tank car leaking chlorine.

"You have many more instances of different chemicals causing you problems than you do radioactive materials," said O'Fallon Chief Michael Ballmann.

"I mean, if you are a gambler, you would go on the odds—and the odds are (in favor of) something like a tractor-trailer or even a normal train derailment of chlorine gas or propane or whatever the case might be."

Major accidents in Missouri, Illinois

Major incidents involving the release of hazardous materials from transport tankers in Illinois and Missouri.

*May 17, 2003—More than 2,500 gallons of hydrochloric acid escaped from an overturned tanker truck on U.S. Highway 61 in Palmyra, Mo. Firefighters temporarily evacuated part of a nearby neighborhood as a precaution that morning. No one was injured, but the acid got into a storm drain that feeds the North River.

*Feb. 9, 2003—About 1,000 residents of Tamaroa, Ill., were forced from their homes for as many as four days after 21 cars of an Illinois Central-Canadian National train derailed in the middle of town. The accident sparked a methanol fire that burned for several hours. Methanol and vinyl chloride escaped from damaged tank cars.

*Aug. 14, 2002—Hundreds of homes and businesses were evacuated and more than 60 people sought medical treatment for various respiratory symptoms after chlorine spewed from a burst hose on a railroad tanker car at DPC Enterprises plant near Festus. Investigators blamed an improperly designed transfer hose.

*Oct. 18, 1998—About 200 people in Pierron, Ill., were evacuated as a precaution after 24 cars of a freight train derailed, including an overturned tank car containing sulfuric acid. The small leak was quickly neutralized, but the accident also temporarily closed Illinois Route 143.

*April 16, 1997—About 18,000 gallons of the chemical methyl isobutyl carbinol spilled from a derailed tank car at the Gateway Western Railway Co. switch yard near Centreville. About 1,000 people were evacuated temporarily. Seventy-eight people went to a nearby hospital complaining of burning eyes or breathing problems.

Reasoning Wheel

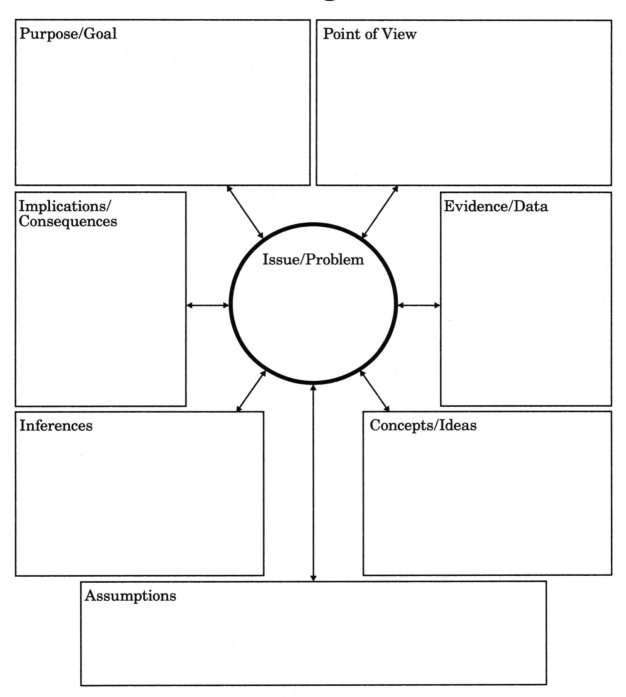

Purpose/Goal

Point of View

Implications/
Consequences

Evidence/Data

Issue/Problem

Inferences

Concepts/Ideas

Assumptions

Adapted from Paul, R. (1992). *Critical thinking: What every person needs to survive in a rapidly changing world.* Sonoma, CA: Foundation for Critical Thinking.

Need to Know Board

What we know . . .	What we need to know . . .	How we can find out . . .

Problem Log Questions

1. Summarize significant points regarding transport regulations for hydrochloric acid.

2. How does this information affect the *Friends of Clear Creek* lawsuit?

3. How does this information affect the problem and our potential solution(s) to date?

Weather Report

The long-range forecast:

Tuesday—High temperature in the 60's, low in the 40's; clear skies, winds from the northwest at 10 miles an hour.

Wednesday—Starting out the same as Tuesday, with highs in the mid-60's, clouding over in the afternoon and winds shifting to the southwest. Chance of showers 20%. Low temperatures in the mid-40's.

Thursday—Cloud cover increasing as the day progresses. Highs in the upper 60's, with winds from the southwest at 20–30 miles an hour. Chance of heavy thunderstorms in the afternoon. Chance of rain 30% in the morning and increasing to 80% in the afternoon.

Friday—Heavy thunderstorms likely, with strong winds from the southwest at 40 miles an hour. Highs in the low 70's; chance of rain 90%.

Problem Log Questions

1. Summarize your solution.

2. How did your group come to this conclusion?

3. Is it more important that the solution be fast, thorough, or inexpensive? Why?

4. What are the criteria that you will use to judge the success of your solution? Briefly describe how each criterion represents an important concern in the development of the solution.

5. Examine your criteria and the solution generated by your group. What weight or significance was given to each criterion? How did each of these work in relation to each other?

6. What sacrifices did you have to make in order to come up with the best possible solution?

7. How did your best possible solution compare to your initial ideas?

Oral Presentation Rubric

Oral Presentation Rubric				
Target Skills	**Novice**	**Developing**	**Proficient**	**Exemplary**
Oral Communication Skills	• The speaker's purpose is unclear.	• The speaker's purpose needs additional clarification.	• The speaker's purpose is clear.	• The speaker's purpose is clear and fully developed.
	• Student maintains little or no eye contact with audience.	• Student sometimes maintains eye contact with audience.	• Student mostly maintains appropriate eye contact with audience.	• Student consistently maintains appropriate eye contact with audience.
	• Student articulation is unclear and difficult to understand.	• Student articulation is generally clear but may not always be correct.	• Student articulates clearly and correctly.	• Student articulates clearly, correctly, and precisely.
	• Student does not use appropriate volume.	• Student is sometimes difficult to hear.	• Student uses appropriate volume most of the time.	• Student uses appropriate volume and considers audience size and room capacity.
	• Distracters in language and body movement overwhelm the presentation.	• Minimal distracters in language and body movement are present.	• Distracters in language and body movement are absent.	• Students uses language and body movement to enhance, not distract from, the presentation.
	• Student word choice is inappropriate and imprecise.	• Student word choice is inappropriate or imprecise.	• Student word choice is appropriate and precise.	• Student word choice is precise and sophisticated.

continued

Oral Presentation Rubric

Target Skills	Novice	Developing	Proficient	Exemplary
Organization Skills	• Little organization is evident in the presentation.	• Limited organization is evident in the presentation.	• Information is presented in an organized sequence.	• Information is presented in an organized and engaging sequence.
	• The presentation shows little evidence of planning.	• The presentation shows some evidence of planning.	• The presentation shows thoughtful planning.	• The presentation shows thoughtful planning and careful construction.
Visual Display Skills	• Visual aid(s) detract from the presentation.	• Visual aid(s) are not appropriately incorporated into the presentation.	• Visual aid(s) complement presentation information.	• Visual aid(s) enhance and reinforce presentation information.
Problem Resolution Skills	• The problem resolution does not consider stakeholder perspectives.	• The problem resolution makes some reference to stakeholder perspectives.	• The problem resolution considers a limited number of stakeholder perspectives.	• The problem resolution considers multiple stakeholder perspectives.
	• There is no connection made between the problem resolution and the information discovered through research.	• Some connection is made between the problem resolution and the information discovered through research.	• Obvious connections are made between the problem resolution and information discovered through research.	• Sophisticated connections are made between the problem resolution and information discovered through research.

Name   Date

Content Post-Assessment

1. a. Draw the pH scale and label its parts.

   b. On your pH scale drawing, show where the following substances belong: highly acidic substances, neutral substances, and highly basic substances.

continued

2. Suppose your teacher gave you a jar filled with a clear liquid and asked you to find its pH. Describe the materials you would need and the method you would use to reach a conclusion.
   Materials:

   Method:

3. How would you change the pH of your clear liquid from acidic to neutral? List the two different ways to do this.
   a.

   b.

4. What safety equipment and precautions would you need to handle your clear liquid safely?
   Equipment:

   Precautions:

Experimental Design
Post-Assessment

Construct a fair test of the following question: *Are bees attracted to diet cola?*

Describe in detail how you would test this question. Be as scientific as you can as you write about your test. Write the steps you would take to find out if bees like diet cola.

Adapted from Fowler, M. (1990). The diet cola test. *Science Scope, 13*(4), 32.

Name _____   Date _____

Systems Post-Assessment

Think of how a swimming pool is a system.

1. List the parts of the system in the spaces provided. Include boundaries, elements, input, and output.

   Boundaries:

   Elements:

   Inputs:

   Outputs:

2. Draw a diagram of the system showing where each of the parts is located.

continued

3. On your diagram, draw lines (in different colors) showing three important interactions between different parts of the pool system. Why is each of these interactions important to the system? Explain your answer.

   a. Interaction #1:

   b. Interaction #2:

   c. Interaction #3: